FRACTIONS IN THE FOREST

THE ADVENTURES OF

JOULE NUMBERS

Copyright © 2015 by
MATH YOU CAN SEE

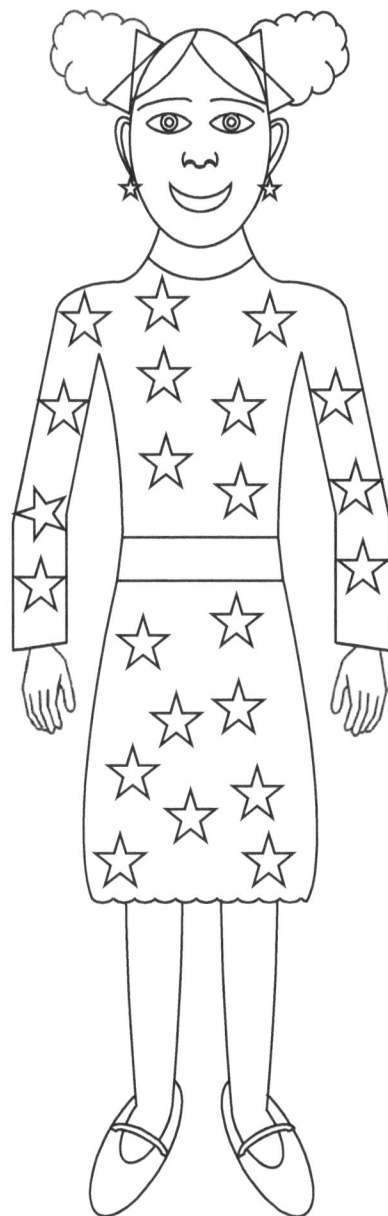

"Hi! I'm Joule Numbers!"

"It's a lovely day outside today. Come take a walk with me in the forest!"

1

As Joule starts out on her walk, she comes across a beautiful flower bed. "I count 24 flowers." says Joule. "One-half of the flowers are blue. One-half of the flowers are pink."

How many flowers are pink?   =   [ ]

How many flowers are blue?   =   [ ]

Workspace:

3

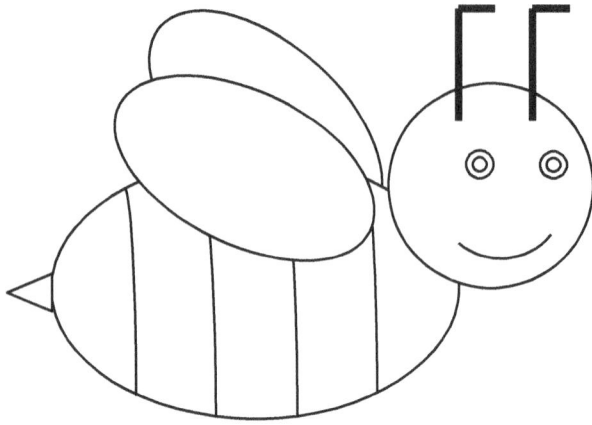

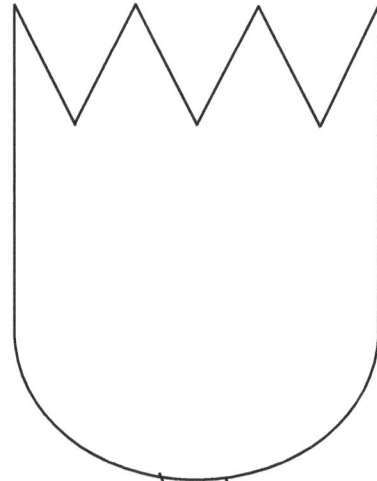

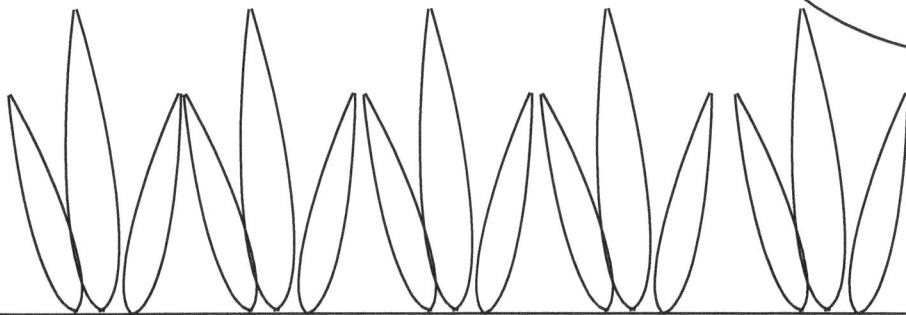

Looking more closely, Joule spots a bumble bee sipping nectar from a blue flower. "Hello Mister Bee.", she says cheerfully. "I see that you have five stripes. Three-fifths of your stripes are black and two-fifths of your stripes are yellow."

4

How many black stripes does the bee have? **=** ☐

How many yellow stripes does the bee have? **=** ☐

Workspace:

At the entrance to the forest, Joule comes across a concession stand. "Oh, look at the pretty colored balloons!" Joule says. Three-eighths of the balloons are green. One-half are red. The rest are blue.

6

How many balloons are red?   =   [ ]

How many balloons are blue?   =   [ ]

How many balloons are green?   =   [ ]

Workspace:

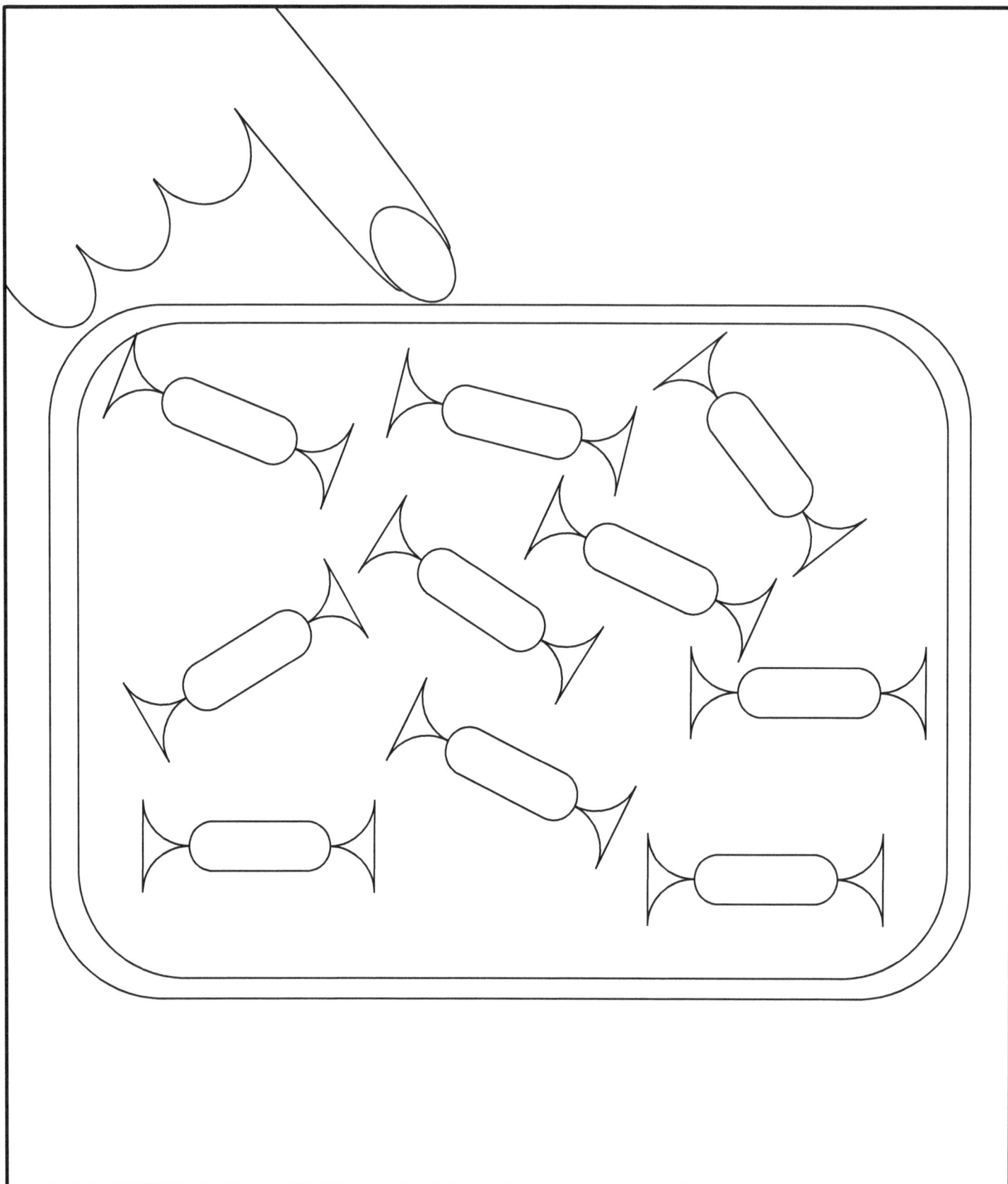

"Good Morning Mister Pi!" says Joule as she inspects the items for sale. Joule spots a tray with ten wrapped candies. One-half of the candies are chocolate and the other half are vanilla. "May I please have two chocolate candies and one vanilla candy?"

8

How many candies are chocolate? = ☐

How many candies are vanilla? = ☐

How many chocolate candies does Mr. Pi now have left after Joule buys two? = ☐

How many vanilla candies does Mr. Pi have left after Joule buys one? = ☐

Workspace:

9

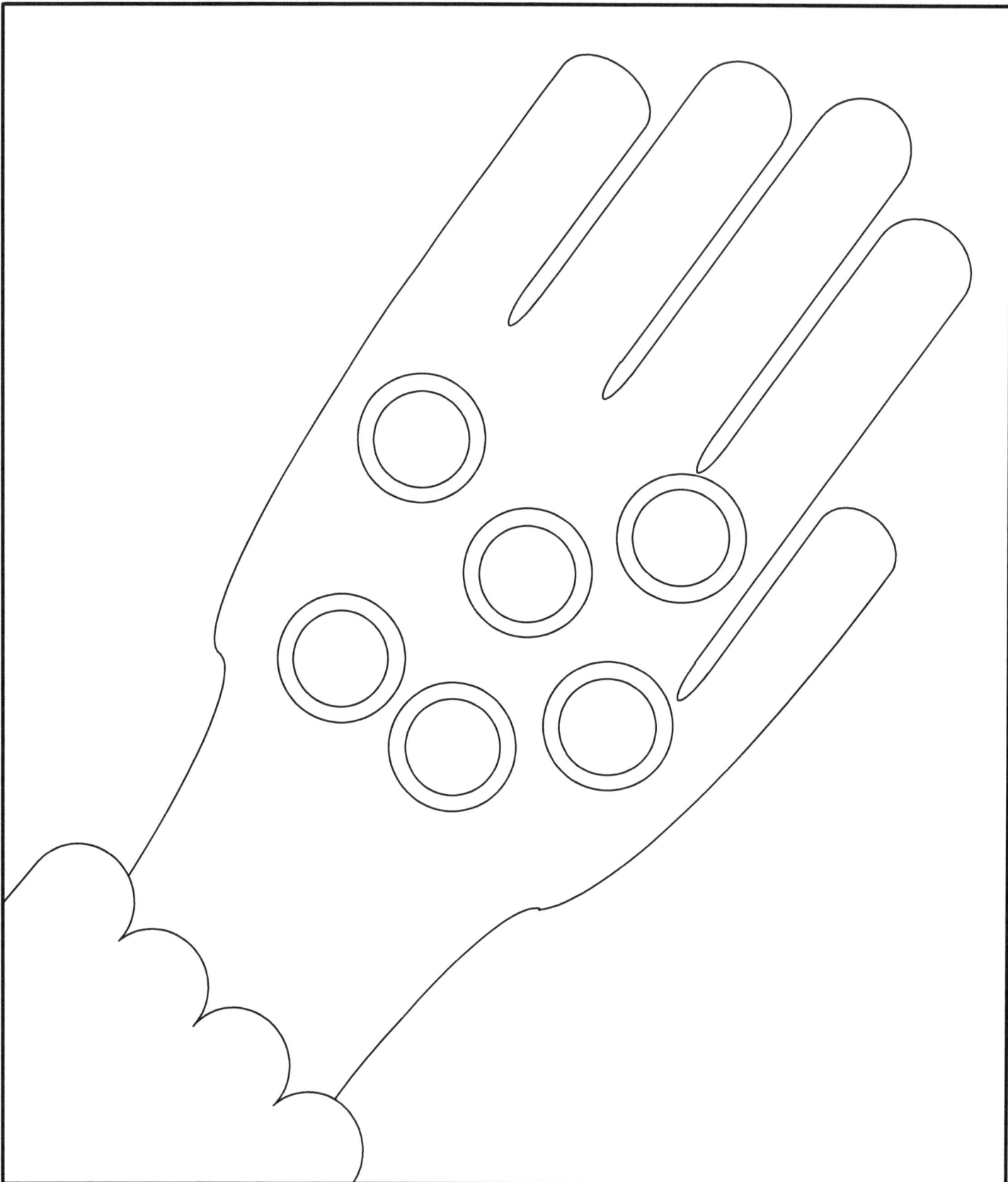

Joule pays for the candies with six coins. One-third of the coins are silver and the rest of the coins are gold. "Have a great day Mr. Pi!" says Joule

How many gold coins does Joule
give to Mr. Pi?

=

How many silver coins does Joule
give to Mr. Pi?

=

Workspace:

11

Joule enters the forest. The forest is thick with tall trees and sweet-smelling flowers. Up ahead she spots an apple tree with 24 apples. The majority of the apples (five eighths) are red. The rest of the apples are light-green.

How many apples on the tree are red?   =   ☐

What fraction of the apples on the tree are light-green?   =   ☐/☐

Workspace:

13

Joule approaches the tree and picks a red apple. *"I will bring this apple home to mom. She will like this."* Says Joule as she tucks the apple in her knapsack.

After Joule picks a red apple, how many red apples remain on the tree?

$=$ ☐

After Joule picks one red apple, what fraction of the apples remaining on the tree are red?

$= \dfrac{}{}$

After Joule picks a red apple, how many light-green apples remain on the tree?

$=$ ☐

Workspace:

15

Joule hears the sound of birds tweeting nearby. She climbs the tree and spots a nest of baby birds. *"Hello Mother Bird! Oh, what cute babies you have!"* Joule counts seven baby birds. Three-sevenths are blue. One is silver and the rest are yellow.

16

How many blue baby birds does
Joule see in the nest?

= ☐

What fraction of the baby birds
that Joule sees are silver?

= ☐/☐

How many yellow baby birds does
Joule see in the nest?

= ☐

Workspace:

Joule feeds the mother bird some bird seed. One-half of the seeds are square and one-half are round. Two of the square seeds are brown. Three of the round seeds are brown. The rest of the bird seeds are yellow. (Hint: Count the seeds in the picture.)

What fraction of all the bird seeds are yellow? ═

How many of the bird seeds are square? ═

What fraction of all the bird seeds are brown? ═

How many of the bird seeds are yellow and round? ═

Workspace:

19

As she trots along, Joule hears a rustling sound nearby. Out from the bushes pops a bobcat. She is orange with 20 markings on her fur. The bobcat has 15 black markings and the rest of the markings are brown. "Here kitty!" says Joule to the bobcat as she pets her head.

What fraction of the bobcat's markings are black?

$=$ $\boxed{—}$

How many of the bobcat's markings are brown?

$=$ $\boxed{}$

What fraction of the bobcat's markings are brown?

$=$ $\boxed{—}$

Workspace:

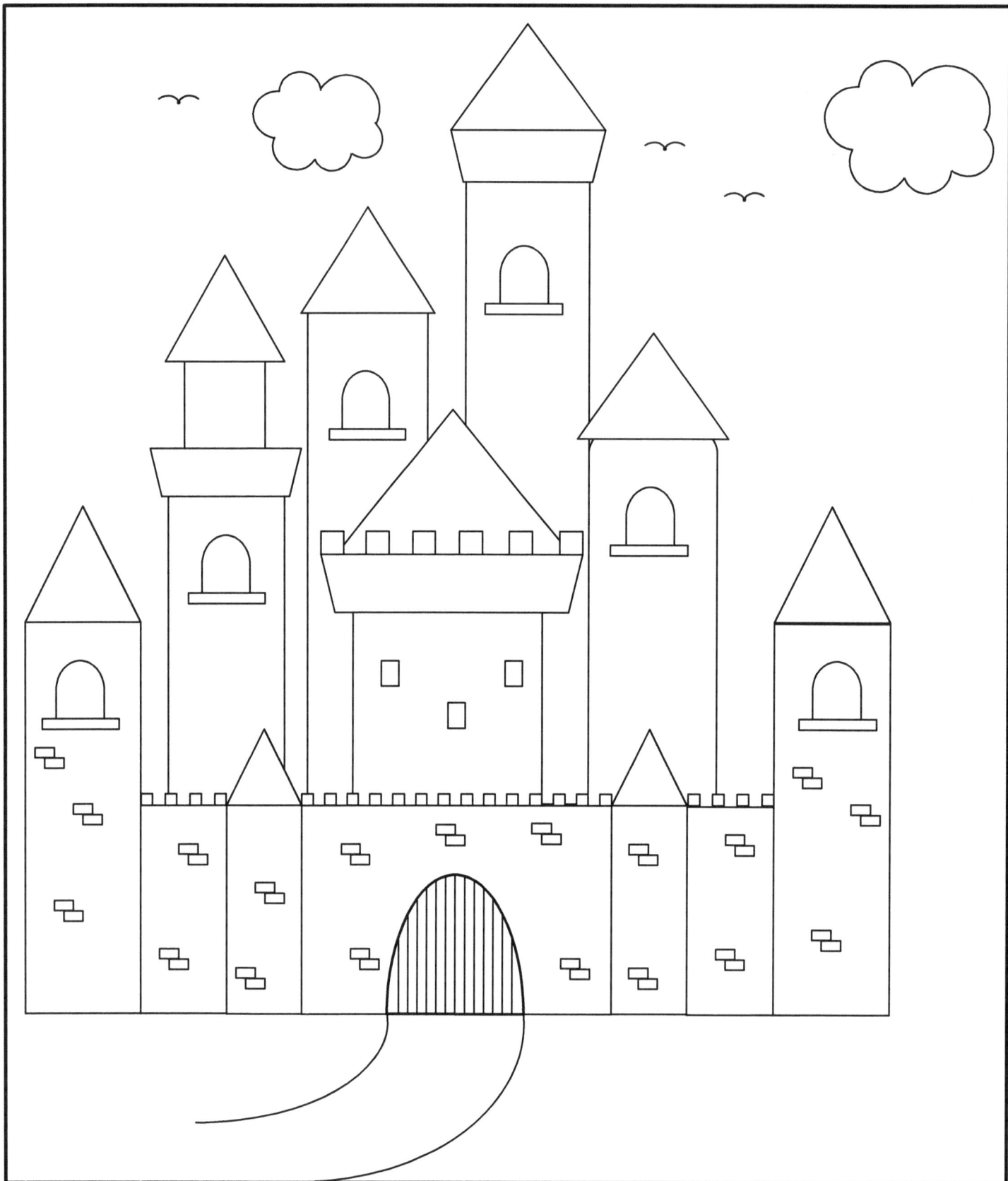

Joule continues on her walk. Up ahead, Joule spots a majestic castle with nine cones. Four-ninths of the cones are grey and the rest are brown.

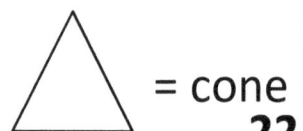

= cone

22

How many of the castle's cones are grey?

$=$

What fraction of the castle's cones are brown?

$=$ ⬜/⬜

Workspace:

23

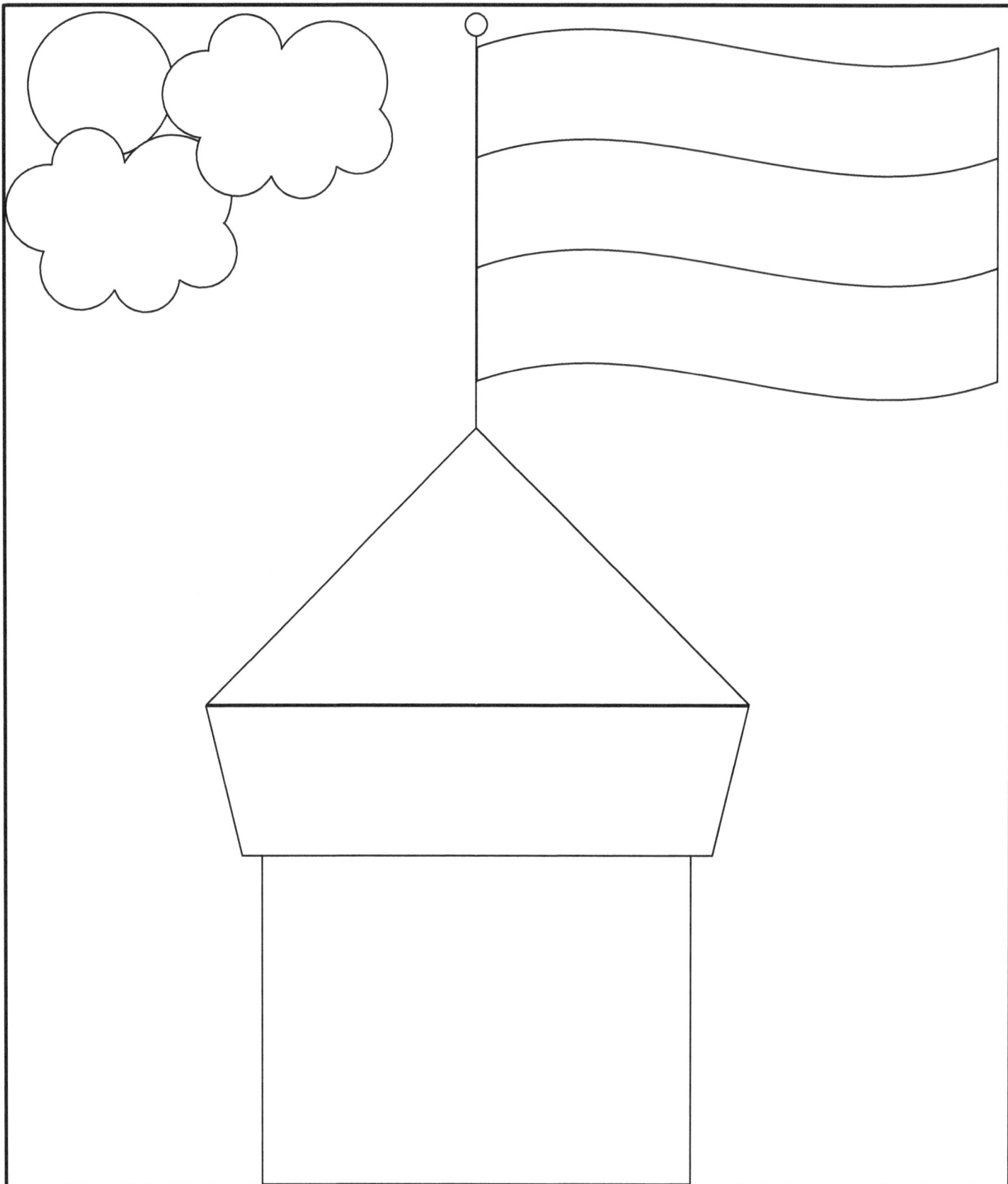

Atop the castle is a large flag flying in the wind. The flag contains three stripes. The outer stripes are green and the inner stripe is silver.

What fraction of the flag's stripes are green?

$=$ $\boxed{\dfrac{}{}}$

What fraction of the flag's stripes are silver?

$=$ $\boxed{\dfrac{}{}}$

Workspace:

25

As Joule approaches the castle, she reaches a bridge that stretches across a stream. Joule stops to look at the school of fish in the stream. She counts 18 fish. One-sixth are orange. One-third are silver. One-ninth are yellow. One-sixth are lavender. The rest are striped green and black.

How many of the fish that Joule sees are orange?　　=　☐

How many of the fish that Joule sees are silver?　　=　☐

How many of the fish that Joule sees are yellow?　　=　☐

How many of the fish that Joule sees are lavender?　　=　☐

What fraction of the fish that Joule sees are green and black striped?　　=　☐

Workspace:

Joule knocks on the door of the castle and opens the door. Once inside, she enters a long hallway with six candles on the walls. Two-thirds of the candles are green and the rest are silver. *"Hello? Is anyone home?"* Joule shouts. Her voice echoes in the empty hallway.

How many green candles are hanging on the wall?   =   [ ]

What fraction of the candles on the wall are silver?   =   [—]

Workspace:

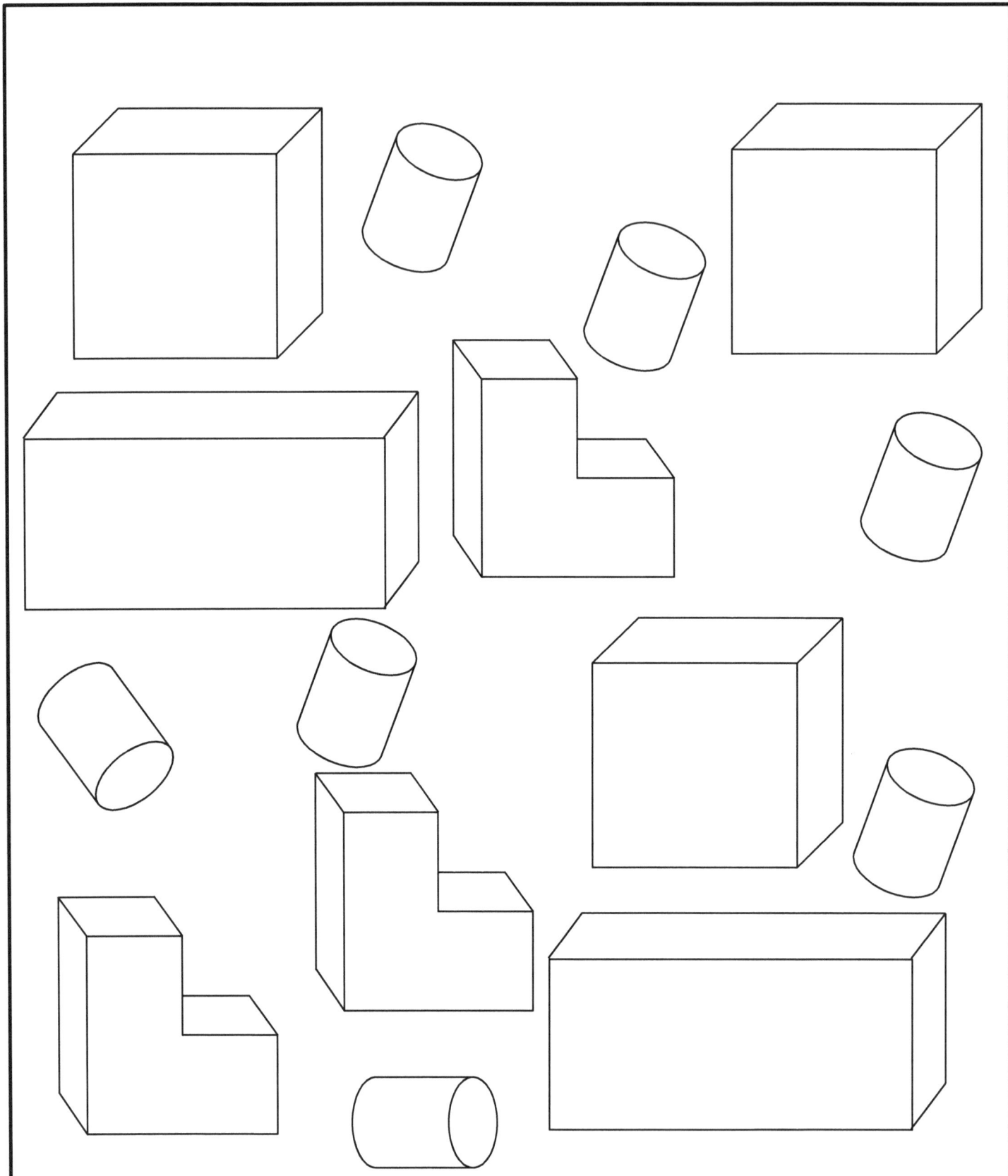

Joule roams the castle and stumbles upon a children's playroom full of toys. On the carpet, she sees 15 toy blocks of various colors. One-fifth are green, two are red, one-third are yellow and the rest are blue. Joule decides to head outside.

How many building blocks are green?

$=$

How many building blocks are yellow?

$=$

What fraction of the building blocks are blue?

$=$

$\dfrac{}{}$

What fraction of the building blocks are red?

$=$

$\dfrac{}{}$

Workspace:

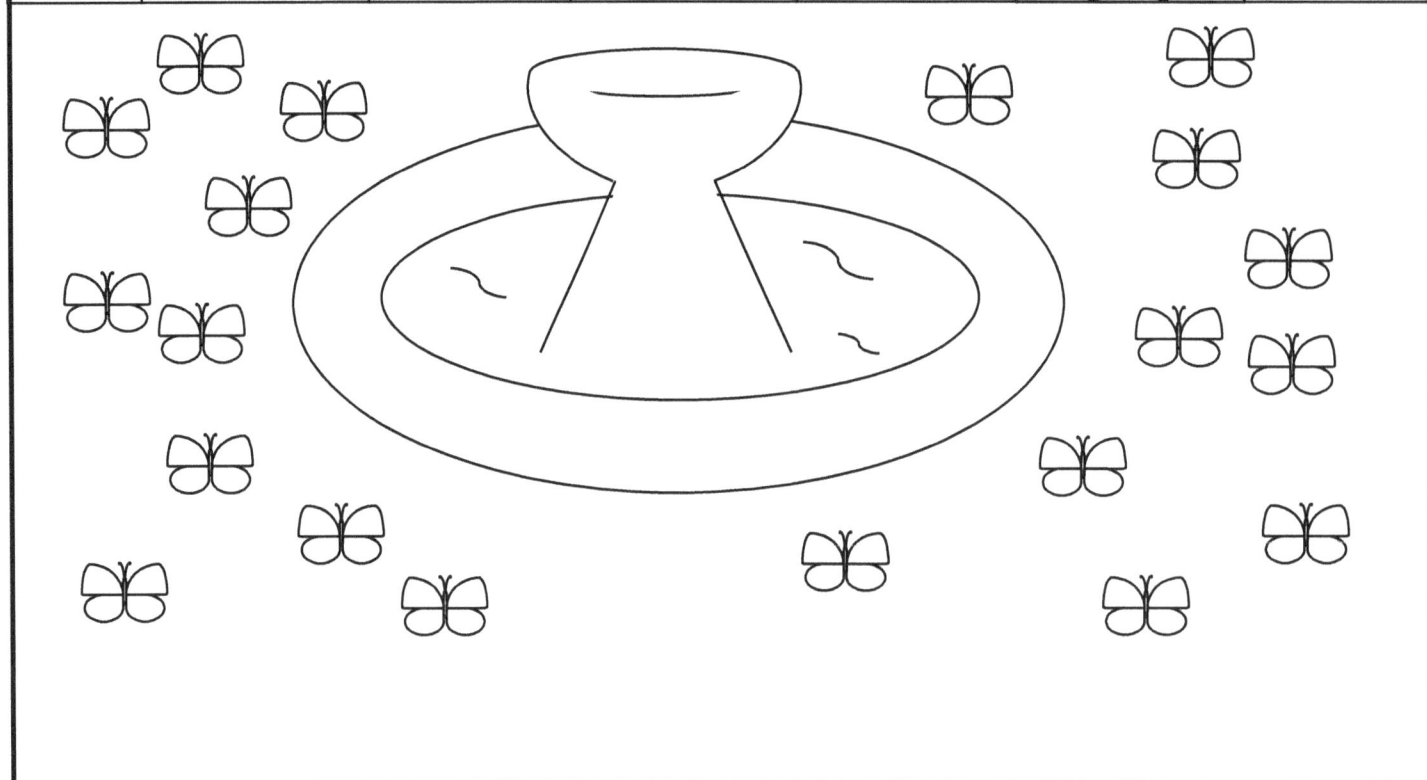

In the private courtyard, Joule sees a beautiful display of 20 butterflies. One-fifth are orange. Three-tenths are turquoise. One-tenth are pink. Two are yellow and the rest are multi-colored.

How many of the butterflies are orange?   ☰   ▢

How many of the butterflies are turquoise?   ☰   ▢

How many of the butterflies are pink?   ☰   ▢

What fraction of the butterflies are yellow?   ☰   ▢̶

What fraction of the butterflies are multi-colored?   ☰   ▢̶

Workspace:

Joule heads back out into the forest. She comes across a large lake and tall mountains in the background. She spots a sailboat with two sails. One-half of the sails are red and the other half is white.

Along her walk, Joule spots an opening to a dark cave. *"I wonder who lives in here."* Joule asks aloud as she peers into the cave. *"Hello? Is anyone home?"* Joule shouts into the cave. Suddenly, in the darkness Joule sees a swarm of. . .

Snakes!!! "YIKES!!" Joule yells as she runs through the forest to escape the snake. The snake is bright yellow with red, brown and green spots. One-third of its spots are green. Two-fifths are brown and the rest of the snake's spots are red. (Hint: Count the snake's spots.)

How many of the snake's spots are green?   ==   [ ]

How many of the snake's spots are brown?   ==   [ ]

What fraction of the snake's spots are red?   ==   [—]

Workspace:

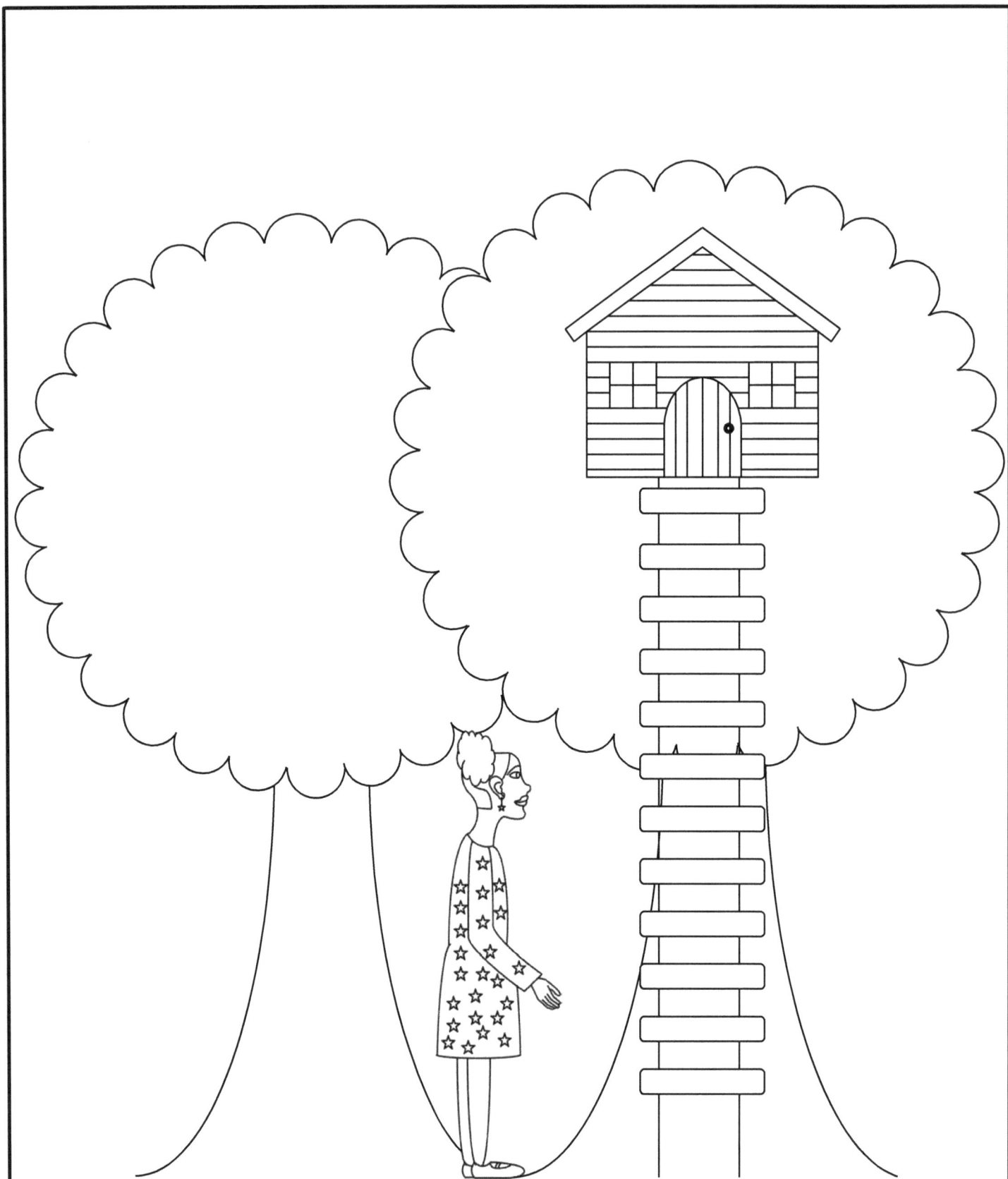

Joule spots a very tall tree house up ahead with a ladder. The ladder has twelve rungs which alternate blue and gold. Joule climbs the ladder to the tree house to escape the snake.

How many gold rungs are on the
ladder that Joule climbs?

=

What fraction of the rungs on the
ladder that Joule climbs is blue?

= ⬚/⬚

Workspace:

Upon entering the tree house, Joule meets a teddy bear who is eating a bowl of berries. There are 18 berries in the bowl. *"Oh Mr. Bear, I am sorry to disturb you."* says Joule. "I *just needed to run from the snake."* The teddy bear offers Joule some berries. One-half of the berries are blue, three are purple and the rest are red. (Hint: Count the berries.)

40

How many of the bear's berries are red?　=　□

How many of the bear's berries are blue?　=　□

What fraction of the bear's berries are purple?　=　□/□

Workspace:

Joule heads back into the forest to return home. She comes across a rose bush with 20 roses. One-half of the roses are pink. One-fifth are red. Three are yellow and the rest are lavender. Joule picks one pink and one yellow rose to bring home to her mom.

42

How many red roses does Joule see on the rosebush? ＝ ☐

How many yellow roses remain on the rosebush after Joule picks one yellow rose? ＝ ☐

How many pink roses remain on the rosebush after Joule picks one pink rose? ＝ ☐

Workspace:

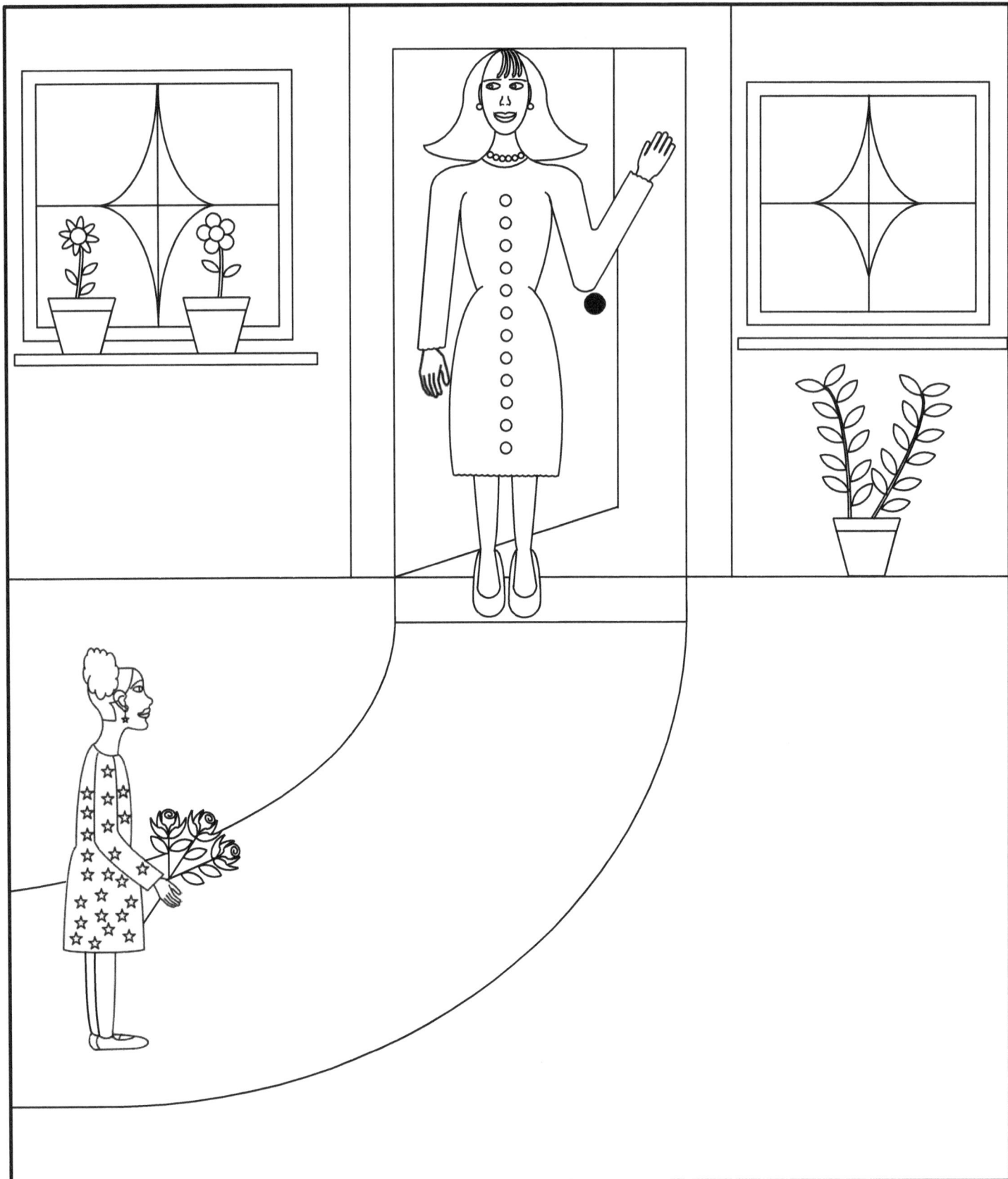

Joule arrives home and presents her mother with the two roses and apple. Joule's mom is wearing a silver dress with 12 buttons on the front. One half of the buttons are white and one-half are pink. "How was your walk in the forest?" Mrs. Numbers asks.

How many pink buttons are on
Mrs. Numbers' dress?   =   ☐

How many white buttons are on
Mrs. Numbers' dress?   =   ☐

Workspace:

45

Once inside, Mrs. Numbers presents Joule with a bowl of three scoops of ice cream. One third strawberry, one-third vanilla and one-third chocolate just for Joule!

Answer Key

Page 3:

    12 pink flowers
    12 blue flowers

Page 5:

    3 black stripes
    2 yellow stripes

Page 7:

    4 red balloons
    1 blue balloon
    3 green balloons

Page 9:

    5 chocolate candies
    5 vanilla candies
    3 chocolate candies left
    4 vanilla candies left

Page 11:

    4 gold coins
    2 silver coins

Page 13:

    15 red apples
    Fraction is 3/8 light-green apples

Page 15:

    14 red apples remain
    Fraction is 14/23 red apples
    9 light-green apples remain

Page 17:

    3 blue baby birds
    Fraction is 1/7 is silver
    3 yellow baby birds

Page 19:

    Fraction is 19/24 yellow
    12 square bird seeds
    Fraction is 5/24 brown
    9 yellow round bird seeds

Page 21:

    Fraction is 15/20 or 3/4
    5 markings are brown
    Fraction is 5/20 or 1/4

Page 23:

    4 cones are grey
    Fraction is 5/9 brown

Page 25:

    Fraction is 2/3 green
    Fraction is 1/3 silver

Page 27:

    3 orange fish
    6 silver fish
    2 yellow fish
    3 lavender fish
    Fraction is 4/18 or 2/9

Page 29:

    4 green candles
    Fraction is 2/6 or 1/3 silver

Page 31:

    3 green blocks
    5 yellow blocks
    Fraction is 5/15 or 1/3 blue
    Fraction is 2/15 red

Page 33:

    4 orange butterflies
    6 turquoise butterflies
    2 pink butterflies
    Fraction is 2/20 or 1/10 yellow
    Fraction is 6/20 or 3/10 multicolored

Page 37:

    5 green spots
    6 brown spots
    Fraction is 4/15 are red spots

Page 39:

    6 gold rungs
    Fraction is 6/12 or 1/2 blue

Page 41:

    6 red berries
    9 blue berries
    Fraction is 3/18 or 1/6 purple berries

Page 43:

    4 red roses
    2 yellow roses remain
    9 pink roses remain

Page 45:

    6 pink buttons
    6 white buttons